User Information

Name

Address

E-Mail

Telephone

Signature

Book Information

Book Number   Start Date

Continued from Book Number

Last Page   End Date

PROJECT INDEX

Page   **Project**

1

2

3

4

5

6

7

8

9

10

11

12

13

14

15

16

17

18

19

20

21

22

23

24

25

26

27

28
29
30
31
32
33
34
35
36
37
38
39
40
41
42
43
44
45
46
47
48
49
50
51
52
53
54
55

56

57

58

59

60

61

62

63

64

65

66

67

68

69

70

71

72

73

74

75

76

77

78

79

80

81

82

83

Page   **Project**

84

85

86

87

88

89

90

91

92

93

94

95

96

97

98

99

100

Continued from Page

Continued in Page

Signature		Date
The Above Confidential Information Was Witnessed and Understood By	Witness Signature	Date

Continued in Page

Signature		Date
The Above Confidential Information Was Witnessed and Understood By	Witness Signature	Date

Continued from Page

Continued in Page

Signature		Date
The Above Confidential Information Was Witnessed and Understood By	Witness Signature	Date

Continued in Page

Signature		Date
The Above Confidential Information Was Witnessed and Understood By	Witness Signature	Date

Continued from Page

Continued in Page

Signature		Date
The Above Confidential Information Was Witnessed and Understood By	Witness Signature	Date

Continued from Page

Continued in Page

Signature		Date
The Above Confidential Information Was Witnessed and Understood By	Witness Signature	Date

Continued from Page

Continued in Page

Signature		Date
The Above Confidential Information Was Witnessed and Understood By	Witness Signature	Date

Continued from Page

Continued in Page

Signature		Date
The Above Confidential Information Was Witnessed and Understood By	Witness Signature	Date

Continued from Page

Continued in Page

Signature		Date
The Above Confidential Information Was Witnessed and Understood By	Witness Signature	Date

Continued from Page

Continued in Page

Signature		Date
The Above Confidential Information Was Witnessed and Understood By	Witness Signature	Date

Continued from Page

Continued in Page

Signature		Date
The Above Confidential Information Was Witnessed and Understood By	Witness Signature	Date

Continued from Page

Continued in Page

Signature		Date
The Above Confidential Information Was Witnessed and Understood By	Witness Signature	Date

Continued from Page

Continued in Page

Signature		Date
The Above Confidential Information Was Witnessed and Understood By	Witness Signature	Date

Continued from Page

Continued in Page

Signature		Date
The Above Confidential Information Was Witnessed and Understood By	Witness Signature	Date

Continued from Page

Continued in Page

Signature		Date
The Above Confidential Information Was Witnessed and Understood By	Witness Signature	Date

Continued from Page

Continued in Page

Signature		Date
The Above Confidential Information Was Witnessed and Understood By	Witness Signature	Date

Continued from Page

Continued in Page

Signature		Date
The Above Confidential Information Was Witnessed and Understood By	Witness Signature	Date

Continued from Page

Continued in Page

Signature		Date
The Above Confidential Information Was Witnessed and Understood By	Witness Signature	Date

Continued from Page

Continued in Page

Signature		Date
The Above Confidential Information Was Witnessed and Understood By	Witness Signature	Date

Continued from Page

Continued in Page

Signature		Date
The Above Confidential Information Was Witnessed and Understood By	Witness Signature	Date

Continued from Page

Continued in Page

Signature		Date
The Above Confidential Information Was Witnessed and Understood By	Witness Signature	Date

Continued from Page

Continued in Page

Signature		Date
The Above Confidential Information Was Witnessed and Understood By	Witness Signature	Date

Continued from Page

Continued in Page

Signature		Date
The Above Confidential Information Was Witnessed and Understood By	Witness Signature	Date

Continued from Page

Continued in Page

Signature	Date
The Above Confidential Information Was Witnessed and Understood By	Witness Signature Date

Continued from Page

Continued in Page

Signature		Date
The Above Confidential Information Was Witnessed and Understood By	Witness Signature	Date

PROJECT

DATE

Continued from Page

Continued in Page

Signature		Date
The Above Confidential Information Was Witnessed and Understood By	Witness Signature	Date

Continued from Page

Continued in Page

Signature		Date
The Above Confidential Information Was Witnessed and Understood By	Witness Signature	Date

Continued from Page

Continued in Page

Signature	Date
The Above Confidential Information Was Witnessed and Understood By Witness Signature	Date

Continued from Page

Continued in Page

Signature		Date
The Above Confidential Information Was Witnessed and Understood By	Witness Signature	Date

Continued from Page

Continued in Page

Signature		Date
The Above Confidential Information Was Witnessed and Understood By	Witness Signature	Date

Continued from Page

Continued in Page

Signature		Date
The Above Confidential Information Was Witnessed and Understood By	Witness Signature	Date

Continued from Page

Continued in Page

Signature		Date
The Above Confidential Information Was Witnessed and Understood By	Witness Signature	Date

Continued from Page

Continued in Page

Signature		Date
The Above Confidential Information Was Witnessed and Understood By	Witness Signature	Date

Continued from Page

Continued in Page

Signature		Date
The Above Confidential Information Was Witnessed and Understood By	Witness Signature	Date

Continued from Page

Continued in Page

Signature		Date
The Above Confidential Information Was Witnessed and Understood By	Witness Signature	Date

Continued in Page

Signature		Date
The Above Confidential Information Was Witnessed and Understood By	Witness Signature	Date

Continued from Page

Continued in Page

Signature		Date
The Above Confidential Information Was Witnessed and Understood By	Witness Signature	Date

Continued from Page

Continued in Page

Signature		Date
The Above Confidential Information Was Witnessed and Understood By	Witness Signature	Date

Continued from Page

Continued in Page

Signature		Date
The Above Confidential Information Was Witnessed and Understood By	Witness Signature	Date

Continued from Page

Continued in Page

Signature		Date
The Above Confidential Information Was Witnessed and Understood By	Witness Signature	Date

Continued from Page

Continued in Page

Signature		Date
The Above Confidential Information Was Witnessed and Understood By	Witness Signature	Date

Continued from Page

Continued in Page

Signature		Date
The Above Confidential Information Was Witnessed and Understood By	Witness Signature	Date

Continued from Page

Continued in Page

Signature		Date
The Above Confidential Information Was Witnessed and Understood By	Witness Signature	Date

Continued in Page

Signature		Date
The Above Confidential Information Was Witnessed and Understood By	Witness Signature	Date

Continued in Page

Signature		Date
The Above Confidential Information Was Witnessed and Understood By	Witness Signature	Date

Continued from Page

Continued in Page

Signature	Date
The Above Confidential Information Was Witnessed and Understood By Witness Signature	Date

Continued from Page

Continued in Page

Signature		Date
The Above Confidential Information Was Witnessed and Understood By	Witness Signature	Date

Continued from Page

Continued in Page

Signature		Date
The Above Confidential Information Was Witnessed and Understood By	Witness Signature	Date

Continued from Page

Continued in Page

Signature		Date
The Above Confidential Information Was Witnessed and Understood By	Witness Signature	Date

Continued from Page

Continued in Page

Signature		Date
The Above Confidential Information Was Witnessed and Understood By	Witness Signature	Date

Continued from Page

Continued in Page

Signature		Date
The Above Confidential Information Was Witnessed and Understood By	Witness Signature	Date

Continued in Page

Signature		Date
The Above Confidential Information Was Witnessed and Understood By	Witness Signature	Date

Continued from Page

Continued in Page

Signature		Date
The Above Confidential Information Was Witnessed and Understood By	Witness Signature	Date

Continued from Page

Continued in Page

Signature		Date
The Above Confidential Information Was Witnessed and Understood By	Witness Signature	Date

Continued from Page

Continued in Page

Signature		Date
The Above Confidential Information Was Witnessed and Understood By	Witness Signature	Date

Continued from Page

Continued in Page

Signature		Date
The Above Confidential Information Was Witnessed and Understood By	Witness Signature	Date

Continued from Page

Continued in Page

Signature		Date
The Above Confidential Information Was Witnessed and Understood By	Witness Signature	Date

Continued from Page

Continued in Page

Signature		Date
The Above Confidential Information Was Witnessed and Understood By	Witness Signature	Date

Continued from Page

Continued in Page

Signature		Date
The Above Confidential Information Was Witnessed and Understood By	Witness Signature	Date

Continued from Page

Continued in Page

Signature		Date
The Above Confidential Information Was Witnessed and Understood By	Witness Signature	Date

PROJECT   DATE

Continued from Page

Continued in Page

Signature		Date
The Above Confidential Information Was Witnessed and Understood By	Witness Signature	Date

Continued from Page

Continued in Page

Signature		Date
The Above Confidential Information Was Witnessed and Understood By	Witness Signature	Date

Continued from Page

Continued in Page

Signature		Date
The Above Confidential Information Was Witnessed and Understood By	Witness Signature	Date

Continued from Page

Continued in Page

Signature		Date
The Above Confidential Information Was Witnessed and Understood By	Witness Signature	Date

Continued from Page

Continued in Page

Signature		Date
The Above Confidential Information Was Witnessed and Understood By	Witness Signature	Date

Continued from Page

Continued in Page

Signature		Date
The Above Confidential Information Was Witnessed and Understood By	Witness Signature	Date

Continued from Page

Continued in Page

Signature		Date
The Above Confidential Information Was Witnessed and Understood By	Witness Signature	Date

Continued in Page

Signature		Date
The Above Confidential Information Was Witnessed and Understood By	Witness Signature	Date

Continued from Page

Continued in Page

Signature		Date
The Above Confidential Information Was Witnessed and Understood By	Witness Signature	Date

Continued from Page

Continued in Page

Signature		Date
The Above Confidential Information Was Witnessed and Understood By	Witness Signature	Date

Continued from Page

Continued in Page

Signature		Date
The Above Confidential Information Was Witnessed and Understood By	Witness Signature	Date

Continued from Page

Continued in Page

Signature		Date
The Above Confidential Information Was Witnessed and Understood By	Witness Signature	Date

Continued from Page

Continued in Page

Signature		Date
The Above Confidential Information Was Witnessed and Understood By	Witness Signature	Date

Continued from Page

Continued in Page

Signature		Date
The Above Confidential Information Was Witnessed and Understood By	Witness Signature	Date

Continued from Page

Continued in Page

Signature		Date
The Above Confidential Information Was Witnessed and Understood By	Witness Signature	Date

Continued from Page

Continued in Page

Signature		Date
The Above Confidential Information Was Witnessed and Understood By	Witness Signature	Date

Continued from Page

Continued in Page

Signature		Date
The Above Confidential Information Was Witnessed and Understood By	Witness Signature	Date

Continued from Page

Continued in Page

Signature		Date
The Above Confidential Information Was Witnessed and Understood By	Witness Signature	Date

Continued from Page

Continued in Page

Signature		Date
The Above Confidential Information Was Witnessed and Understood By	Witness Signature	Date

Continued from Page

Continued in Page

Signature		Date
The Above Confidential Information Was Witnessed and Understood By	Witness Signature	Date

Continued from Page

Continued in Page

Signature		Date
The Above Confidential Information Was Witnessed and Understood By	Witness Signature	Date

Continued from Page

Continued in Page

Signature	Date
The Above Confidential Information Was Witnessed and Understood By Witness Signature	Date

Continued from Page

Continued in Page

Signature		Date
The Above Confidential Information Was Witnessed and Understood By	Witness Signature	Date

Continued from Page

Continued in Page

Signature		Date
The Above Confidential Information Was Witnessed and Understood By	Witness Signature	Date

Continued from Page

Continued in Page

Signature		Date
The Above Confidential Information Was Witnessed and Understood By	Witness Signature	Date

Continued from Page

Continued in Page

Signature		Date
The Above Confidential Information Was Witnessed and Understood By	Witness Signature	Date

Continued from Page

Continued in Page

Signature		Date
The Above Confidential Information Was Witnessed and Understood By	Witness Signature	Date

Continued from Page

Continued in Page

Signature		Date
The Above Confidential Information Was Witnessed and Understood By	Witness Signature	Date

PROJECT   **DATE**

Continued from Page

Continued in Page

Signature		Date
The Above Confidential Information Was Witnessed and Understood By	Witness Signature	Date

Continued from Page

Continued in Page

Signature		Date
The Above Confidential Information Was Witnessed and Understood By	Witness Signature	Date

Continued from Page

Continued in Page

Signature		Date
The Above Confidential Information Was Witnessed and Understood By	Witness Signature	Date

Continued in Page

Signature		Date
The Above Confidential Information Was Witnessed and Understood By	Witness Signature	Date

Continued from Page

Continued in Page

Signature		Date
The Above Confidential Information Was Witnessed and Understood By	Witness Signature	Date

Continued from Page

Continued in Page

Signature		Date
The Above Confidential Information Was Witnessed and Understood By	Witness Signature	Date

Continued from Page

Continued in Page

Signature		Date
The Above Confidential Information Was Witnessed and Understood By	Witness Signature	Date

Continued from Page

Continued in Page

Signature		Date
The Above Confidential Information Was Witnessed and Understood By	Witness Signature	Date

Continued from Page

Continued in Page

Signature		Date
The Above Confidential Information Was Witnessed and Understood By	Witness Signature	Date

Continued from Page

Continued in Page

Signature		Date
The Above Confidential Information Was Witnessed and Understood By	Witness Signature	Date

Continued from Page

Continued in Page

Signature		Date
The Above Confidential Information Was Witnessed and Understood By	Witness Signature	Date

Continued from Page

Continued in Page

Signature		Date
The Above Confidential Information Was Witnessed and Understood By	Witness Signature	Date

Made in the USA
Las Vegas, NV
29 December 2023

83677503R00059